Comparing Bugs

A World of Bugs

Charlotte Guillain

www.heinemannraintree.com
Visit our website to find out more information about Heinemann-Raintree books.

To order:

☎ Phone 888-454-2279

🖳 Visit www.heinemannraintree.com to browse our catalog and order online.

© 2012 Heinemann Library
an imprint of Capstone Global Library, LLC
Chicago, Illinois

Customer Service: 888-454-2279
Visit our website at www.heinemannraintree.com

Edited by Rebecca Rissman, Daniel Nunn, and Harriet Milles
Designed by Joanna Hinton-Malivoire
Picture research by Elizabeth Alexander
Originated by Capstone Global Library Ltd.
Production by Victoria Fitzgerald
Printed and bound in China by Leo Paper Products Ltd

15 14 13 12 11
10 9 8 7 6 5 4 3 2 1

Library of Congress Cataloging-in-Publication Data
Guillain, Charlotte.
 A world of bugs / Charlotte Guillain.
   p. cm.
 Includes bibliographical references and index.
 ISBN 978-1-4329-5505-2 (hc)—ISBN 978-1-4329-5506-9 (pb) 1.
Insects—Juvenile literature. I. Title.
 QL467.2.G858 2011
 595.7—dc22   2010045845

Acknowledgments
The author and publishers are grateful to the following for permission to reproduce copyright material: Alamy **p. 12** (© blickwinkel): Corbis **p. 21 left** (© Hans Pfletschinger/Science Faction); FLPA **p. 19** (© Michael Durham/Minden Pictures); iStockphoto **p. 18 right** (© Dawn Hudson); Shutterstock **pp. 4 left, 22 bee** (© Daniel Hebert), **4 right, 22 butterfly** (© Leighton Photography & Imaging), **5 left, 22 pill bug** (© Joseph Calev), **5 right, 9** (© Audrey Snider-Bell), **6, 22 dragonfly** (© iliuta goean), **7 left** (© David Dohnal), **7 right** (© basel101658), **8** (© orionmystery@flickr), **10** (© Dariusz Majgier), **11** (© Wong Hock weng), **13 left & right** (© Cathy Keifer), **14, 22 beetle** (© argonaut), **15 left** (© Chris Mole), **15 right** (© Patrick Power), **16** (© Ariel Bravy), **17** (© Four Oaks), **18 left** (© Vinicius Tupinamba), **20** (© Gregory Guivarch), **21 right, 22 water strider** (© mjf99).

Front cover photograph of a banana spider and grasshopper on a leaf reproduced with permission of Shutterstock (© Cathy Keifer). Back cover photograph of a dung beetle in South Africa rolling a ball of dung reproduced with permission of Shutterstock (© Four Oaks).

We would like to thank Michael Bright for his invaluable help in the preparation of this book.

Every effort has been made to contact copyright holders of any material reproduced in this book. Any omissions will be rectified in subsequent printings if notice is given to the publisher.

Some words appear in bold, **like this**. You can find out what they mean in "Words to Know" on page 23.

Contents

About this series

Books in this series invite readers into the exciting world of bugs! Use this book to stimulate discussion about what bugs are, where they can be found, how they move and eat, and how different they can look.

What Are Bugs?

bee

butterfly

Bugs are small living things. They do not have **backbones**. There are different types of bugs. Some bugs are **insects**. Bees and butterflies are insects.

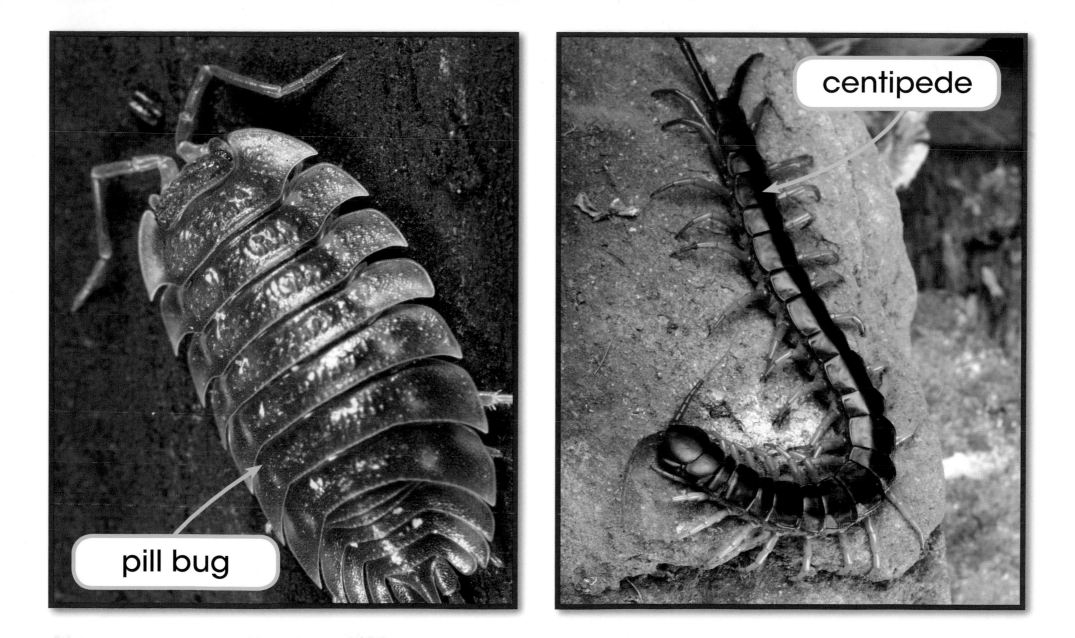

pill bug

centipede

Not all bugs are insects. Spiders, pill bugs, centipedes, and worms are other types of bugs.

Bugs' Bodies

wings

dragonfly

Insects have three main body parts, and six legs. Many insects, such as moths and dragonflies, have wings. Insects' eyes are made of many little eyes next to each other.

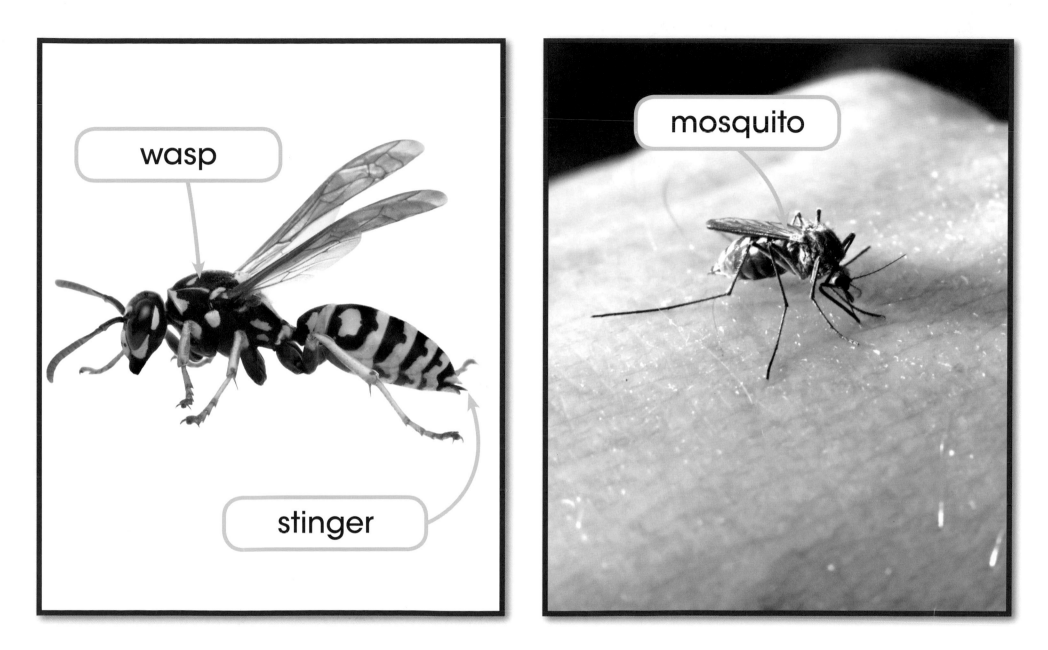

wasp

stinger

mosquito

Some insects **sting** or bite. Female bees, wasps, and ants have stingers. Lice, fleas, and mosquitoes use their mouths to bite, and suck up liquid.

spider

Spiders have eight legs and two body parts. They have six or eight eyes. They have **fangs** to bite. Spiders do not have wings.

antennae

centipede

Centipedes and millipedes have long bodies with many different sections. They have **antennae** and many legs. Centipedes can bite with their claws.

Growing Bugs

stink bug

eggs

Most bugs lay eggs. Centipedes lay eggs in the soil. Butterflies, ladybugs, and stink bugs lay eggs on leaves or grass. Mosquitoes lay eggs in water. Spiders wrap their eggs in silk.

eggs

young bugs

Some bugs look like their parents when they **hatch**.
Young millipedes and centipedes look like small
adults. Young grasshoppers and spiders look like
small adults, too.

larvae

Not all bug babies look like adults. Some eggs **hatch** into **larvae**. Larvae change into **pupae** as they grow. Then they become adults.

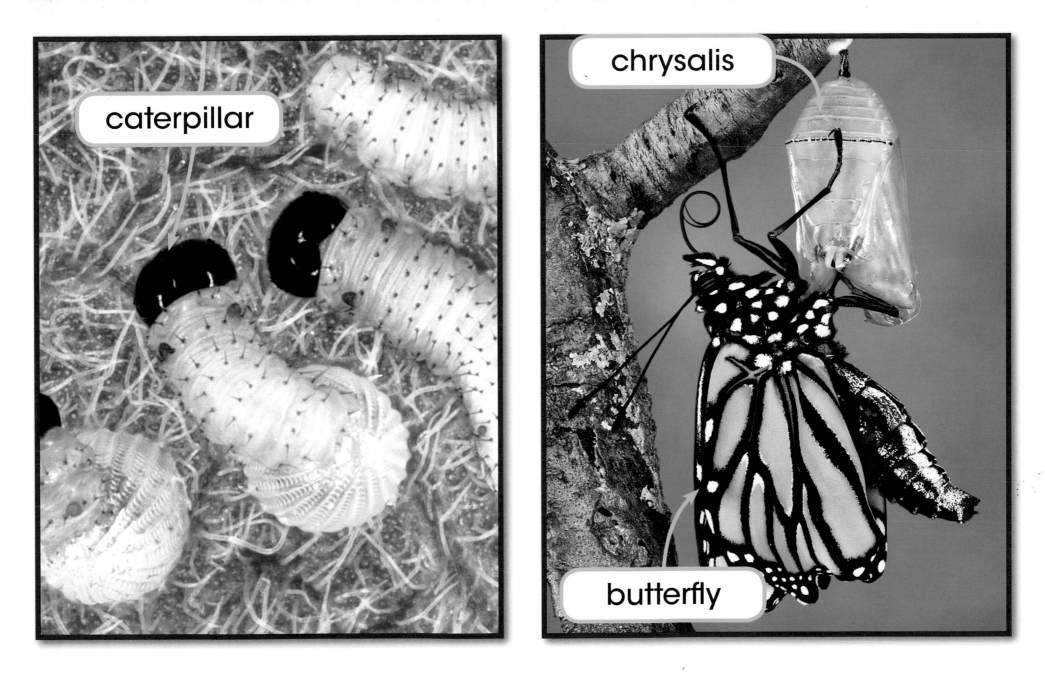

caterpillar

chrysalis

butterfly

A caterpillar is a larva. The caterpillar grows and changes into a pupa called a **chrysalis**. Then it changes into a butterfly inside the chrysalis.

Bugs' Homes

beetle

Earthworms, ants, and some centipedes live underground. Many beetles and pill bugs live under logs or dead leaves.

web

nest

spider

wasp

Spiders make webs from silk. Bees build nests from **wax**. Some wasps chew things up to make a nest. Other bugs live in plants and trees.

Bugs' Food

butterfly

Many bugs get food from plants. Some bugs eat leaves. Bees and butterflies feed on **nectar** from flowers. Other bugs eat fruit or seeds.

dung beetle

Some bugs eat other bugs. Mosquitoes and ticks suck blood. Dung beetles and some flies eat animal manure. Some big spiders eat mice or birds!

Moving Bugs

worm

beetle

Many bugs run or crawl. Caterpillars, worms, and snails move quite slowly. Ground beetles and spiders move more quickly.

grasshopper

Some bugs can jump. Some **insects**, such as grasshoppers, crickets, and fleas, have strong legs to help them jump. Some spiders can jump, too.

hawkmoth

Many **insects** can fly. Hawkmoths fly very quickly. Large beetles fly slowly. Spiders, pill bugs, centipedes, and millipedes cannot fly. Some spiders can float on the wind.

diving beetle

water strider

Some insects can swim. Diving beetles paddle with their back legs. Water striders move on the top of the water. Water spiders swim underwater.

How Big are Bugs?

These pictures show you how big some of the bugs in this book are.

Words to Know

antenna feeler on an insect's head. Antennae can be used to feel, taste, and smell.

backbone part of a skeleton that goes from the tail to the head

chrysalis where a caterpillar turns into a butterfly

fangs sharp teeth that can be used to give a poisonous bite. Spiders have fangs.

hatch born from an egg

insect type of bug with three body parts and six legs

larvae the young of some types of insect

nectar sweet liquid that is made by flowers

pupae stage in an insect's life cycle between larva and adult

stinger part of an insect's body that can give venom

wax sticky substance that bees use to build their nests. Some candles are made from wax.

Index

Note to Parents and Teachers

Before reading

Make a list of bugs with the children. Try to include insects, arachnids (e.g., spiders), crustacea (e.g., wood lice), myriapods (e.g., centipedes and millipedes) and earthworms, slugs, and snails. Ask them what body parts they think each bug has. Do they think bugs have the same senses as us? Do they know where different bugs live, or what they eat?

After reading

- Between spring and late summer you could go on a bug hunt. Divide the children into groups and give each group a plastic pot, a paintbrush and a magnifying glass. Go out into the school grounds and look under stones and leaves for bugs. Show the children how to gently put any bugs they find into the plastic pot using the paint brush and then look at them using a magnifying glass. Emphasize how important it is to treat living creatures carefully and to put them back where they were found. Ask the children to try to identify the bugs they find and look at the body parts they have. Tell them to record where they found each bug. Share their findings at the end of the hunt.

- Get a butterfly kit for your classroom to watch how caterpillars grow and change into butterflies. Ask the children to make a diary recording how the caterpillars change.

- Get a class wormery to observe what compost worms eat. Discuss how these worms help us to recycle waste and help the environment.